THE BEST OF LEMONS

THE BEST OF LEMONS

4447
Published in the USA 1995 by JG Press
Distributed by World Publications, Inc.

Printed and bound in Italy
ISBN 1-57215-044-0

The JG Press imprint is a trademark of JG Press, Inc.
455 Somerset Avenue
North Dighton, MA 02764

CONTENTS

INTRODUCTION

Lemons can be used in every kind of dish, from savory soups to desserts, as both flavoring and garnish. And slices of lemon may well have appeared before the start of the meal itself, floating in pre-dinner drinks. Their usefulness does not end here. Lemon juice is used to acidulate water in which slices of food such as apples are soaked to prevent them turning brown when exposed to the air, or it can be brushed directly over cut surfaces that would otherwise discolor rapidly. On pale meats such as chicken or rabbit, lemon juice acts as a mild, harmless bleaching agent. It also makes a fine alternative to mild wine or cider vinegar, far preferable to the harsh white variety, for mayonnaises and salad dressings. Finally, always have a squeezed-out lemon half on a soap dish by the kitchen sink. Rub it on your hands to kill any odors left after handling strong-smelling foods like fish or garlic.

The original home of the lemon is thought to be south-eastern Asia. From there it gradually came west until in about 1,000 AD the Arabs introduced it into the Mediterranean region. To begin with, it was only used medicinally, but gradually its culinary importance took precedence. Today, large quantities of lemons are grown all over the world.

Choosing lemons is easy once you know what to look for. Like all citrus fruit, a juicy specimen will feel heavy for its size. The deeper yellow its color, the less acid it is likely to be. Examine the skin, too. Thin-skinned fruit is usually finely textured. A thick skin often looks lumpy and is deeply pock-marked. Avoid hard-skinned lemons (which are probably old and may be rather dry inside) and any which have soft, spongy patches (they are probably already decaying inside and will soon turn moldy all over).

Lemon juice and grated rind are constantly used in the kitchen, often in very small amounts. When only a few drops of juice are called for, there is no need to cut a whole lemon in half. Simply use a thick needle or sharp skewer to puncture the skin at one end in several places, and squeeze the whole fruit until drops of juice run out. If you need to squeeze out a whole fresh lemon, its juice will run more freely if you first soften the fruit by rolling it over a hard surface under the palm of your hand.

Never waste the zest, even if it is not to be used immediately. Grate it off *before* halving and squeezing the fruit and store it in a tightly covered container in the freezer. Use it straightaway in its frozen state when you need a pinch for a recipe.

Invariably, however, there will be times when a lemon half, both rind and juice, is left over. It can be stored for two or three days in the door of the refrigerator, the cut surface tightly sealed with plastic wrap. (A whole lemon whose skin has been grated off should be kept in the same way.)

Finally, a good way of making the most of lemon chunks that might otherwise go to waste is to cut them in slices or wedges, arange them in one layer on a cookie sheet (so they are not touching) and freeze them hard. The pieces of lemon can then be tied up in a plastic bag ready to add, still frozen, to drinks, instead of fresh lemon and ice cubes. Lemons are almost as valuable medicinally as they are in cooking, and are also used cosmetically in a number of ways.

AVGOLEMONO

½ cup rice
5 cups well-flavored chicken stock
2 egg yolks
¼ cup lemon juice
Thin slices of lemon and finely chopped parsley, to garnish

Rinse the rice in a colander under the cold tap. Cover with cold water and leave to soak for 15 minutes.

Meanwhile, bring chicken stock to simmering point. Drain rice thoroughly, stir it into the stock and cook gently until tender, about 15 minutes. Remove pan from the heat.

In a bowl, beat egg yolks and lemon juice together thoroughly. Gradually beat in a ladleful of the hot stock and slowly pour the mixture back into the remaining soup, stirring constantly.

Reheat gently, stirring and taking great care that the soup does not come to the boil, or the egg yolks will curdle. Taste and flavor with more lemon juice, if liked. The soup should be very sharp and lemony.

Serve immediately, each portion garnished with a thin slice of lemon and a sprinkling of chopped parsley.

Serves 6

VEAL IN LEMON PARSLEY SAUCE

6 thin slices veal scallops
Seasoned flour
¼ cup butter
1½ tablespoons olive oil
Juice of 1 large lemon
2–3 tabelspoons finely chopped parsley

Pound each scallop with a dampened rolling pin and cut it into small squares. Dust on both sides with seasoned flour.

In a large frying pan, heat the olive oil and half the butter, and brown veal quickly on both sides. Remove veal from the pan. Pour off remaining fat and add the rest of the butter. When this has melted, stir in the lemon juice and parsley. Stir over low heat until well mixed.

Return veal to the pan. Mix well until thoroughly hot and coated with buttery juices, and serve immediately.

Serves 6

BAKED FISH WITH LEMON SOUR CREAM SAUCE

Butter
1 thin-skinned lemon
1 mild, medium onion
1½ lbs small white fish fillets: flounder, haddock, red snapper, etc.
Salt and freshly ground black pepper
1 cup sour cream (see below)
¼ teaspoon sweet paprika
1 teaspoon coarse whole-grain mustard (eg Moûtarde de Meaux)

Grease a large, flat baking dish lightly with butter. Cut lemon horizontally into thin slices, discarding end pieces which are mostly rind and pith. Pick out pips as you come across them and discard these as well. Cut onion into paper-thin slices. Arrange lemon and onion rings in baking dish in alternating, overlapping rows. Lay fish fillets on top in a single layer. Sprinkle them lightly all over with salt and freshly ground black pepper, and dot with a few flakes of butter. Cover dish with a sheet of aluminum foil and bake fish in a moderately hot oven 400°F, for 20 minutes, or until it can be flaked with a fork.

While fish is in the oven, switch on broiler so that it will be thoroughly hot when needed. Blend sour cream with paprika, mustard and a pinch of salt.

Uncover fish. Pour sour cream mixture over the entire surface, spreading it out evenly. Immediately broil fish about 3 inches from source of heat until surface is lightly colored. Serve immediately.

Serves 4

Note: If you prefer a lighter dish, substitute half plain yogurt and half créme fraiche, lightly beaten, or use yogurt on its own, beaten until smooth.

SEVICHE

1½ lbs fillets of sole or pompano skinned (see method)
⅔ cup fresh lemon or lime juice

Dressing
½ cup olive oil
2 medium tomatoes, peeled, seeded and diced
2 canned hot green chilli peppers, rinsed, seeded and chopped (but see method)
1 small mild onion, finely chopped
2 tablespoons finely chopped parsley
½ teaspoon chopped oregano
Salt and freshly ground black pepper

Ask your fishmarket to skin fish fillets for you or do it yourself as follows. Lay a fillet on its skin with the pointed (tail) end towards you. With a sharp knife cut fish from skin at the pointed end, going far enough to allow you to get a good grip on the skin. Then turn fillet over and pull skin sharply so it rips off in one piece, holding fish down with the other hand.

Cut skinned fillets into thin strips and lay them out evenly in a deep glass or porcelain dish. Pour over lemon or lime juice, adding some more if necessary to cover fish completely. Cover dish and leave to marinate in the bottom of the refrigerator for 3-4 hours, or until strips have lost their translucent, raw look and have turned opaque. Turn fish strips several times while they are marinating to ensure they are constantly coated in juice.

Prepare dressing. Assemble all the ingredients in a bowl and stir gently until thoroughly mixed. If you are not afraid of quite a fiery sauce, leave some of the seeds in the peppers. On the other hand, if hot green chillis are not available, a dash or two of Tabasco may be substituted.

Shortly before serving, drain fish thoroughly. Mix gently with dressing and chill lightly until ready to serve.

Serves 6

Note: A garnish of sweet, mild onion rings, slices of peeled and stoned avocado or whole green olives is sometimes used.

ARTICHOKE SALAD

3 large juicy lemons
3 teaspoons salt
10–15 canned artichoke hearts
Scant 1 cup whole roasted almonds
1 teaspoon honey
2 tablespoons olive oil
Juice of 1 large lemon

Wash the lemons under cold running water, but do not peel or cut. Place them in a saucepan. Add the salt and enough water to cover, and bring to a boil over a medium heat. Continue to cook until the fruit starts to soften, then remove from the heat and drain. Leave the lemons to cool. When the lemons are cold, slice thinly and throw into a chilled salad bowl. Halve the artichoke hearts and add them to the lemons. Add the roasted almonds, the honey and the oil. Leave in a cold place until ready to serve (do not prepare more than 2 hours in advance). Just before serving, toss and sprinkle the lemon juice over the salad.

Serves 4

LEMON SOUFFLÉ

1½ tablespoons unflavored powdered gelatin
3 eggs, separated
¾ cup sugar
Finely grated rind and strained juice of 3 large lemons
⅔ cup heavy cream

Sprinkle gelatin over 3 or 4 tablespoons cold water in a cup and leave to soften for a few minutes. When mixture turns hard and stiff, stand cup in a small pan of very hot water and stir until gelatin has dissolved and liquid is quite clear. Allow to cool to room temperature. In a heatproof bowl, work egg yolks with sugar and beat in lemon rind and juice. Fit bowl snugly over a pan of simmering water, and beat over gentle heat until mixture is thick and fluffy. Remove bowl from pan and continue beating for a few minutes longer to cool it. Then slowly whisk in dissolved gelatin.

Whisk egg whites to a firm snow. In another bowl, carefully whisk cream until thick and floppy but not stiff. Fold cream into lemon mixture, followed by egg whites. Spoon into a serving bowl and chill until firm.

Serves 6

LEMON PANCAKES

1 cup all-purpose flour
Pinch of salt
1 egg
⅔ cup milk
1 tablespoon oil
Brown sugar
Lemon juice
Butter

Start by preparing pancake batter. Sift flour and salt into a bowl, and make a well in the center. Break in the egg. Dilute milk with ⅔ cup water. Working with a large wooden spoon, break up the egg and gradually work it into the flour, slowly adding diluted milk as you do so. When batter is smooth and free of lumps, beat in oil. Put aside to 'rest' for 1 hour before frying pancakes.

When ready to fry pancakes, heat your pancake (or omelet) pan (about 7 inches in diameter), and quickly brush surface with a butter paper or a wad of kitchen paper dipped in oil. Pour in just enough batter to coat base of pan, pouring back any excess left over after a thin layer has set on surface of pan. (Scrape off 'trail' left on side of pan by pancake batter when pouring excess out.)

When pancake is lightly colored on the underside, loosen edges all round with a spatula and flip it over. Color on the other side, then turn out on to an upturned soup plate and cover with a folded dish towel to keep pancake soft and moist. Stack pancakes on top of one another as you prepare them.

When time comes to serve pancakes, scatter each one with brown sugar and sprinkle with lemon juice, leaving a border of about 1 inch clear around sides. Fold pancake in four.

In a large, heavy frying pan, melt a large knob of butter. When foaming, arrange a single layer of folded pancakes in pan and fry steadily until crisp and golden. Turn pancakes over and fry on other side, adding a little more butter as needed.

Serve hot, sprinkled with a little more sugar and lemon juice if liked.

Serves 4–6

CITRUS BAKE

3 eggs, separated
Finely grated rind and juice of 2 large lemons
1 cup sugar
Butter
6 tablespoons flour
1¼ cup milk
Pinch of salt

Beat egg yolks and lemon rind together until light. Gradually add sugar, beating vigorously. If mixture becomes too stiff to absorb all the sugar, thin it with some of the lemon juice, then beat in remaining sugar, followed by remaining lemon juice, and continue to beat until white and fluffy. Beat in a tablespoon of melted butter.

Sift the flour on to the lemon mixture, a tablespoon at a time, and gently stir it in. Gradually stir in the milk. In a large bowl, beat the egg whites with a pinch of salt until they form stiff peaks.

Using a large metal spoon, lightly and gradually fold lemon mixture into the beaten egg whites.
Pour mixture into a large buttered baking dish and bake in a moderate oven, 350°F, for 35–40 minutes, or until pudding feels springy to the touch and is a rich golden color.
This is at its best eaten warm.

Serves 4–6

SUSSEX POND PUDDING

2 cups flour
2 teaspoons double-action baking powder
Pinch of salt
½ cup finely chopped beef suet
Milk and water, mixed
Butter, for pudding basin

Filling
½ cup butter
⅔ cup brown sugar
1 large, thin-skinned lemon, frozen and thawed

To make dough, sift flour, baking powder and salt into a bowl, and stir in suet with a fork until thoroughly mixed. Then add enough milk diluted half and half with water to make a softish dough.

Butter a medium-sized pudding basin or deep baking dish. Take a quarter of the dough and roll it out on a lightly floured board to make a 'lid' for the basin. Put aside. Roll out remaining dough and line the basin with it.

To make filling, slice the butter thickly and place half of it in a bowl, together with half of the sugar. Prick the lemon all over with a sharp skewer, and place it on top. Cover with the remaining butter and sugar. Fit the pastry lid on top and press well to seal it to pastry on sides of basin.

Cover top of basin loosely with aluminum foil or a double thickness of wax paper and tie firmly around sides with string. Place basin on a trivet in a large pan. Pour in boiling water to come halfway up sides of basin, cover tightly with a lid and boil gently for 3½–4 hours, topping up pan with more boiling water as necessary.

To serve, turn pudding out on to a deepish serving dish to catch the buttery sweet sauce that flows out of it. Serve each portion with a piece of lemon.

Serves 6–8

Note: The pudding will take an hour to cook in a pressure cooker at 15 pounds pressure.

LEMON BARS

One 9 in. pie shell or large rectangular pastry base, pre-baked, made with pie dough using
2 cups flour

large, thin-skinned lemons
4 cups sugar
2 cups ground almonds

Peel the lemons, scraping off white pith, and slice them horizontally, discarding pips. In a deep, wide dish, sprinkle the slices with 1 cup of the sugar and put aside for at least 1 hour to allow sugar to dissolve.

In a heavy pan, mix together the almonds, the rest of the sugar, the syrup from the lemons and 3 tablespoons water. Stir over low heat until it becomes a smooth paste.

Arrange lemon slices neatly on the pastry base. Cover them evenly with the almond mixture and place in a slow oven, 325°F, for 20 minutes, or until top feels dry. Cool slightly before cutting into bars with a sharp knife.

Serves 8

LEMON ICE CREAM

Juice and finely grated rind of 1 medium lemon
Scant 1 cup sugar
1¾ cup light cream

Blend lemon juice, grated lemon rind and sugar together until they are syrupy. Gradually blend or beat in cream. Taste and add a little more lemon juice or sugar if necessary, bearing in mind that freezing will weaken flavors.

Pour mixture into a freezing tray, cover with a lid or a sheet of foil and freeze until mixture has solidified around sides and is thick and mushy in the middle.

Scrape out into a bowl and beat thoroughly to break down any large ice crystals. When mixture is smooth and creamy, quickly pour back into freezing containers, cover and freeze until solid.

Transfer to main cabinet of refrigerator for about 1 hour to soften slightly before serving.

Serves 4

MERINGUE MILE-HIGH PIE

A 9 inch pastry shell

Filling
⅓ cup cornstarch
Generous pinch of salt
About ¾ cup sugar
⅔ cup fresh lemon juice
3 egg yolks
2 rounded teaspoons butter

Meringue
3 egg whites
Pinch of salt
¾ cup sugar
1 teaspoon lemon juice of vanilla essence

Leave the shell in its pie pan (preferably one with a removable base).

In a heavy pan, stir cornstarch with salt and ½ cup sugar until well mixed. Gradually stir in 1¾ cup hot water and bring to the boil over high heat, stirring constantly. Lower heat and simmer for 3-4 minutes stirring frequently, until mixture is thick and smooth and no longer tastes floury. Remove pan from heat.

Beat egg yolks with remaining sugar. Slowly beat in several tablespoons of the hot cornstarch sauce, then pour it all back into the pan beating or stirring vigorously with a wooden spoon.

Place the pan over low heat and cook gently for a few minutes longer, stirring constantly, until sauce thickens again. Remove from the heat, beat in grated lemon rind and butter, and cool to lukewarm. Taste and add a little more lemon or sugar if you like.

Make a meringue. Whisk the egg whites with a pinch of salt until they form very soft, floppy peaks. Add half of the sugar, a tablespoon at a time, whisking vigorously until mixture is stiff and glossy. Finally, fold in the remaining sugar and the lemon juice or vanilla.

Pour the cooled lemon filling into the shell. Cover with meringue, spreading it right to the pastry rim so that the filling is sealed in (this will help prevent the meringue shrinking back while it is in the oven). Flick up the surface of the meringue in rough peaks with the tip of a knife blade.

Bake the pie in a slow to moderate oven, 325–350°F for 15–20 minutes, or until the surface of the meringue feels firm to the touch and the tips of the peaks are lightly colored. Serve lukewarm.

Serves 6–8

FRESH LEMON JELLY

Finely pared rind of 2 lemons
One 2 in. cinnamon stick
2–3 cloves
¼ cup powdered unflavored gelatin
¾ cup sugar
1¼ cup fresh lemon juice

In a pan, cover the lemon rind, cinnamon stick and cloves with 1¼ cups cold water. Slowly bring to boiling point, remove from heat and leave to 'infuse', covered, for 15 minutes.

Meanwhile, sprinkle gelatin over 6 tablespoons water taken from 2½ cups. Leave to soften for 10 minutes, then place container in a bowl of hot water and stir until gelatin has dissolved and liquid is quite clear.

Stir into infused lemon liquid, which should still be warm.

Add sugar and stir until dissolved. Finally, stir in lemon juice and the rest of the water. Strain into a large jelly mold and when quite cold, chill in the refrigerator until firmly set. Turn out and serve.

Serves 6

SYLLABUB

1 orange
1 lemon
3 tablespoons sugar
⅔ cup heavy cream
½ wine glass dry sherry
Peeled pistachio nuts, to decorate

Finely grate the rinds and squeeze the juice of the orange and lemon into a large bowl. Add the sugar and stir until dissolved. Add the cream and sherry, and whisk slowly (preferably with a hand whisk) until syllabub is thick.

Pile in a glass serving bowl or in individual, tall, stemmed glasses (wine glasses, for example). Spike with pistachio nuts and serve accompanied by a dish of shortbread fingers.

Serves 4

LEMONY CHEESECAKE

½ pound smooth cottage or Ricotta cheese
3 eggs, separated
½ cup sugar
⅔ cup sour cream
Juice and finely grated rind of 2 large lemons
½ teaspoon vanilla essence
¼ cup flour
1 tablespoon candied orange peel, diced
1 tablespoon raisins

Beat the cheese until smooth and creamy by hand or in a processor. Beat in the egg yolks one at a time. Then, beating vigorously, gradually add the sugar, followed by the sour cream, lemon juice and vanilla essence. Sift flour over the surface of the mixture, folding it in gently, together with the lemon rind, candied orange peel and raisins.

In a large bowl, beat egg whites until stiff but not dry. Using a large metal spoon, carefully fold cheese mixture *into* the beaten egg whites. Spoon mixture into a well-buttered, 8 inch loose-sided cake pan. (Or use an ordinary deep cake pan, butter it, line the bottom with a disc of wax paper and butter this as well.)

Bake the cheesecake in a slow oven, 325°F, for 1 hour, or until it feels firm to the touch and is lightly colored on top. Switch off the oven and without opening the oven door, allow the cheesecake to stand for an hour. Then open the oven door and cool it for a further 30 minutes. Remove the cheesecake from the oven, let it stand until quite cold and chill it thoroughly before unmolding.

Makes 8 wedges

LEMON OR LIME SHERBET

⅔ cup fresh lemon or lime juice
1½ cups sugar
A few drops of green food coloring
1 egg white

Squeeze lemon or lime juice, pour it into a jug and chill until ready to use it. You can also use a mixture of the two flavors.

In a pan, dissolve sugar in 3 cups water and bring to boiling point. Simmer for 2 or 3 minutes. Remove from heat and allow to cool, stirring occasionally. Stir in chilled juice and tint syrup green with a few drops of food coloring. Pour into a shallow plastic box which has a lid or into ice cube trays, cover with the lid or with a sheet of foil and freeze until slushy, like wet snow.

When ready to proceed, first beat egg white to soft peak stage – it should not be too stiff. Turn the frozen syrup out into a bowl and beat it thoroughly to break down any large ice crystals. Then beat in egg white, pour back into freezing containers, and freeze until mushy again.

Sorbet can now be frozen until semi-firm and served, but if there is time, its texture will be even smoother if you turn it out into a bowl and beat it up vigorously once more before re-freezing it.

Serves 6

HERB BUTTER

½ cup butter, softened
Juice and rind of 1 lemon
1 tablespoon finely chopped fresh parsley
1 teaspoon finely chopped fresh chives
½ teaspoon each finely chopped fresh basil and fresh chervil

Mash all the ingredients together in a bowl, or use an electric blender. Mix thoroughly. Use immediately or place in sealed container and store in the refrigerator until needed.

LEMON CURD

4 juicy, thin-skinned lemons
4 eggs
½ cup butter, softened
1½ cups sugar

Finely grate the lemon rinds, taking care to get rid of any bitter white pith. Squeeze lemon juice and put it through a strainer. Beat eggs until well mixed and foamy.

In the top of your double boiler, combine lemon rind and juice, softened butter and sugar, and stir occasionally over simmering water until butter has melted. Strain in the beaten eggs, stirring vigorously to blend all the ingredients thoroughly. Continue to stir over simmering water until the curd is smooth, thick and creamy, about 20 minutes. It will be very hot, but do not let it boil, or the eggs will curdle.

Pour the curd into small, hot, clean jars and allow to cool before covering as you would jam and storing in the refrigerator. It will keep for a maximum of 8 weeks.

Fills about 3 jars

LEMON MARMALADE

12 juicy thin-skinned lemons
5 cups sugar

With a potato peeler, peel the lemons very thinly so that you take only the colored zest. Cut it into thin, long slivers.

Squeeze the lemons. Collect all the pips, tie them up in muslin bag and put them in a preserving pan with 5 cups cold water. Leave to soak for at least 2 hours.

When ready to make the marmalade add the lemon juice and shredded zest to the pan with the soaking pips, bring to the boil and boil briskly for 20 minutes. Stir in the sugar and cook gently, stirring until it has dissolved. Then raise the heat again and boil until setting point is reached.

Remove the bag of pips. Pour marmalade into hot, sterilized jars and cover as usual.

Fills about 4 jars

FRESH AND FROSTY LEMONADE

2 cups sugar
3 oranges
6 large lemons

In a heavy pan, dissolve sugar in 2½ cups water over low heat. Bring to the boil, then remove from heat and leave to cool.

Meanwhile, grate enough rind from the oranges to make 2 tablespoons. Add it to the cooling syrup together with the squeezed juice of the oranges and lemons, and mix well.

Leave for 1 hour.

Pour through a muslin-lined strainer. Pour into a stoppered bottle or jug and store in the refrigerator. It will kccp for a few days.

LEMON FRESHENER

Lemon juice is a powerful natural bleach, and soothes the skin as well as whitening it. Instead of throwing out old lemon halves, rub them on roughened and discolored elbows. A lemon juice rinse is a very effective hair lightener, especially if you dry your hair in the sun. It also makes a good setting lotion.

LEMON FACE CREAM

These lotions are easy to make and take advantage of the soothing and whitening properties of lemons.

For a toning and cleansing lotion, roughly chop one or two lemons, put them in an enamel or earthenware pan, press to extract the juice, cover with milk and bring very gently to the boil. Turn off the heat, allow to cool, strain, and add a little glycerine – just enough to give a pleasant consistency.

The following lotion will bleach dingy skin, sunburn and freckles, and will whiten and soften rough hands. Simply shake together roughly equal parts of lemon juice, glycerine and rosewater. Some people prefer less glycerine.

REMEDIES

For a sore throat, squeeze the juice of a lemon, or chop it roughly and simmer it in water for about 20 minutes to make a lemony liquid. In either case, when mixed with honey, hot water, and whiskey or vodka, it makes a soothing toddy that helps fight infection.

Lemon juice is a natural antibiotic, tonic, diuretic and restorative. A teaspoonful in a glass of warm water, taken first thing in the morning, promotes regularity, cleanses and generally tones up the system, and helps digestion. It is not usually recommended in cases of gout or rheumatism because of its acidity, but an infusion of the leaves can be substituted. This also helps to induce sleep, and has been used to bring down fevers in illnesses such as typhoid.

Lemon peel is also good for fevers as it helps induce sweating.

OVEN TEMPERATURES

Very low	250°F
Low	275°F
Very slow	300°F
Slow	325°F
Moderate	350°F
	375°F
Moderately hot	400°F
Fairly	425°F
Hot	450°F

NOTES: